Small
WILD CATS

By Sommer Estrada

Small Wild Cats

Written by Sommer Estrada

ISBN: 978-1-7343-2147-0 (Paperback)
Imprint: Sommer Estrada
Printed in the United States of America
Kindle Direct Publishing
Book #4 First Edition: August 2020

Dedicated,

to cat lovers,

worldwide.

The Sand Cat

- The Sand cat is also known as the sand dune cat.
- The sand cat is found in the desert regions of Africa and Asia.
- They find shelter in abandoned burrows and may also dig their own.
- They can survive extreme temperatures from 20 to 128 degrees farenheit.

- The height, from shoulder to ground, is 9 to 12 inches. The body and head length is 15 to 20 inches.
- They can have 1 to 8 kittens at one time. The kittens become independent of their mother between 5 and 8 months.
- Their lifespan is about 13 years.

The Pallas Kedisi

- The Pallas Kedisi is native to central Asia, from Western Iran to Western China.
- They prefer cold, rocky, and grassland areas.
- They are found in elevations up to 16,500 feet.
- The Pallas has the longest and most dense fur of any wildcat family, which helps them withstand temperatures of negative 50 degrees farenheight.
- They are not fast runners. They hunt by stalking and ambushing their prey.
- They prey upon rodents, birds, and insects.

- Their head and body length is 19 to 26 inches.
- They have a stocky posture with a unique feline face, due to a shorter jaw with fewer teeth.
- Their eyes have round pupils instead of a vertical shape.
- Their average lifespan is about 12 years.
- They have between 2 and 8 kittens at one time.
- Its scientific name means "ugly-eared." Most don't agree.

The Rusty-Spotted Cat

- The Rusty-spotted cat lives in India, Sri Lanka, and Nepal.
- They live in bamboo forests, wooded grasslands, dry forests, and rocky hillslopes.
- Their diet includes rodents, birds, insects, and frogs. They are also known for stealing domestic chickens and ducks.

- The Rusty-spotted cat is the smallest species of cats in the world; much smaller than a house cat.
- The body length is 12 to 19 inches, and a height of about 8 inches.
- They weigh between 2 and 5 pounds.
- The lifespan of the Rusty-spotted cat is 12 to 14 years. The longest living Rusty-spotted cat lived 18 years in captivity.

The Black Footed Cat

- The Black footed cat is named for the solid, black color on the bottom of their paws.
- The Black footed cat is the most successful feline hunter. They catch 60% of thier prey. They have a high metabolism; with an average, nightly kill every 50 mintues.
- They prey on rodents, reptiles, insects, birds, rabbits, and eggs.
- The Black footed cat lives in southern areas of Africa; mostly semi-deserts and savanna grasslands.

- The Black footed cat is similar in size to the Rusty-spotted cat.
- It is also one of the smallest wild cats; with a weight of about 3 to 5 pounds.
- They have 1 to 3 kittens, and give birth in burrow type dens.
- The Black footed cat lives up to 8 years in the wild, but can live 16 years in captivity.

The Ocelot

- The fur of the Ocelot is considered one of the most beautiful in the cat family. Sadly, hunting has diminished their numbers. They are a protected species.
- The Ocelot can live 7 to 10 years in the wild, but up to 20 years in captivity.
- They are similar in size to a Bobcat.
- The Ocelot usually has 2 kittens at a time.

- The Ocelot is native to southwestern United States, such as Texas and Arizona. They are found in Mexico, Central and South America.
- The diet of the Ocelot includes frogs, snakes, iguanas, birds, rodents, rabbits, monkeys, fish, and crabs.
- The Ocelot is a very picky eater, and will pluck feathers and remove fur before eating their meal.
- The ocelot is a great swimmer.
- The biggest threat to the Ocelot is the loss of its habitat, due to farming, logging, and growth of civilization.

The Serval

- The Serval lives in Africa, and is nicknamed the "giraffe cat," due to its long legs. Their legs are longer than any cat, relative to its body size.
- They live in the savannas of northern and southern Africa.
- They hunt birds, reptiles, insects, fish, rodents and they use their huge ears to help zone in on their meal.
- They can leap up to 10 feet into the air to catch their prey.
- Some breeders have bred the Serval with the domestic cat and named the breed "Savannah." But the Savannah tends to maintain its wild nature, and may not make a good house cat.

- The "giraffe cat" has a shoulder height between 21 and 26 inches, and a length of 30 to 40 inches.
- The Serval can weigh between 20 to 40 pounds.
- They have the largest ears of any cat species. The ears are black on the back with a white spot, (see back cover.)
- The Serval has been recorded to have lived 22 years in captivity.

The Scottish Wildcat

- The Scottish Wildcat closely resembles the look and size of a domestic cat.
- They live to about 7 years in the wild and 15 years in captivity.
- Females can have between 1 to 8 kittens, and tend to leave them around 6 months.
- They prey on rabbits, voles, rats, birds, fish and insects.

- The Scottish Wildcat lives in northern and eastern Scotland.
- The Scottish Wildcat is the rarest wildcat, and is near extinction.
- They are endangered due to hunting, loss of habitat, and breeding with domestic cats; which diminishes the gene pool.
- The Scottish Wildcat is fearless of water; they usually will swim across rivers to investigate other areas.

The Asian Leopard Cat

- The Asian Leopard cat is found in 21 countries in Asia; the widest territory of the wild, Asian felines.
- They cover semi deserts, low land tropical rainforests, and moist, dry forests in the Himalayan foothills at elevations of 3,000 feet to 14,000 feet.
- They prey on birds, rodents, insects, bats, rabbits, snakes, fish, and amphibians.

- The size of the Asian Leopard varies due to an estimated twelve subspecies.
- They usually have 1 to 4 kittens.
- The Asian Leopard cat usually pairs for life.
- The Asian Leopard cat has been bred with the domestic cat to create the "Bengal" cat breed.

The Caracal

- The Caracal lives in Africa, Middle East, and parts of India.
- Its diet consists of rodents, hares, birds, small monkeys, gazelles, and antelope.
- The Caracal is nicknamed the African Lynx or Persian Lynx, but not related to the Lynx family.
- They have long hair tufts on the ends of their ears. The tufts help them hear prey better, and are also used to communicate with other Caracals.

- The Caracal is known for their jumping ability. They have very strong hindlegs, and can jump 12 feet high.
- They store their food in trees to save for a later meal.
- The Caracal weighs between 25 to 42 pounds, and is considered the heaviest of the small wildcats.
- They have sandy, reddish brown fur.
- The Caracal lives up to 16 years in captivity, and around 12 years in the wild.

The Lynx

- There are four different subspecies of cats in the lynx family; the Canadian Lynx, Bobcat, Eurasian Lynx, and Iberian Lynx.
- The Lynx has a typical cat diet, but its diet also includes large prey, such as goats, sheep, turkeys, and carabou.
- The lynx can weigh between 18 to 40 pounds.

- The Lynx is known for its short, black tipped tail.
- The hips on a Lynx are higher than the shoulders when compared to a Bobcat.
- They can have long ear tufts, long legs and large paws that serve as natural snow shoes. Their large furry feet naturally spread, giving them more surface space to walk across snow.
- The Lynx is found in Europe, Asia and North America.

The Bobcat

- The Bobcat is part of the Lynx family.
- The Bobcat has a short, striped tail with a bicolored tip, black on the upper side and white below.
- The Bobcat has smaller ear tufts, and a smaller body than their other Lynx family members.
- The Bobcat weighs between 20 and 26 pounds, while his other Lynx family members can weigh up to 40 pounds.

- The Bobcat inhabits parts of the United States, and areas of Mexico.
- The males cover a territory of 30 square miles; the female has a territory of only 5 square miles.
- The Bobcat's diet consists of rodents, rabbits, squirrels, birds, and sometimes deer.
- The Bobcat is named after their short, bobbed tail; averaging 5 to 7 inches.
- Bobcats maintain large numbers in the wild, and are not on the endangered species list.

Made in the USA
Monee, IL
04 November 2026

67374639R00017